R. S.

## A Fovre-Fovld Meditation, of the Foure Last Things

viz. ... 1. Houre of Death. 2. Day of Iudgement. 3. Paines of Hell. 4. Loyes of

Heauen

R. S.

**A Fovre-Fovld Meditation, of the Foure Last Things**
*viz. ... 1. Houre of Death. 2. Day of Iudgement. 3. Paines of Hell. 4. Loyes of Heauen*

ISBN/EAN: 9783337208684

Printed in Europe, USA, Canada, Australia, Japan

Cover: Foto ©berggeist007 / pixelio.de

More available books at **www.hansebooks.com**

# ISHAM REPRINTS.

*THE ISHAM REPRINTS.*

*No. 4.*

# A FOVRE-FOVLD MEDITATION.

BY R. S.

1606.

# A Foure=Fould Meditation,

## Of the foure last things :

### viz.

| | | | |
|---|---|---|---|
| 1. | | | Houre of Death. |
| 2. | of the | | Day of Iudgement. |
| 3. | | | Paines of Hell. |
| 4. | | | Ioyes of Heauen. |

### Shewing the estate of the Elect and Reprobate :

COMPOSED IN A DIUINE POEME

## By R. S.

*The author of S. Peters complaint.*

## [ROBERT SOUTHWELL, S.J.]

Imprinted at London by *G. Eld:* for *Francis Burton.*
1606.

WITH A BIBLIOGRAPHICAL PREFACE

## By CHARLES EDMONDS;

EDITOR OF THE "ISHAM SHAKESPEARE ; " "BASILICON DORON OF K. JAMES I. ; "
"HAKE'S NEWES OUT OF POWLES CHURCHYARDE ; "
"THE POETRY OF THE ANTI-JACOBIN, BY THE RT. HON. G. CANNING, THE
RT. HON. J. HOOKHAM FRERE, G. ELLIS, W. GIFFORD, ETC. ; "
"THE PYTCHLEY HUNT, PAST AND PRESENT, BY H. O. NETHERCOTE."

PUBLISHED BY

# ELKIN MATHEWS,

VIGO STREET, LONDON.

MDCCCXCV.

CHISWICK PRESS:—CHARLES WHITTINGHAM AND CO.
TOOKS COURT, CHANCERY LANE, LONDON.

# A BIBLIOGRAPHICAL NOTE BY THE
# DISCOVERER AND EDITOR.

AS the merits of Southwell, both as a Poet and a Martyr, have been continually eulogized by Catholics and Proteſtants alike, it is unneceſſary to dilate upon them here. My intention is, therefore, to addreſs myſelf only to the diſcovery and ſubſequent adventures of the intereſting Tractate, now for the firſt time ſubmitted to the notice of the public.

It was one amongſt many of the valuable works of Old Engliſh Poetry and Proſe of the Elizabethan and Jacobean ages which I diſcovered at Lamport Hall in September, 1867, and the circumſtances under which it was brought to light, and its author's identity proved, are ſo uncommon that they might form a chapter in a Romance of Bibliography. The facts are theſe :

After the iſſue of Nos. 1 and 2 of the "Iſham Reprints," which were the hitherto-unknown edition of Shakeſpeare's "Venus and Adonis" of 1599, and Hake's rare "Newes out of Powles Churchyarde" of 1579, the next volume of which I recommended the publication was "A Foure-fould Meditation of the Foure Laſt Thinges ; compoſed in a Divine Poeme. By R. S., the author of S. Peters Complaint," London, 1606, if the miſſing portion of the

poem could be found, for I had only a flight fragment containing the firft eight leaves alone ; but
thefe were precious, as in addition to the firft 35
ftanzas, they gave, not only a Dedication by W. H.[1]
(himfelf a literary difcoverer) in thefe ftriking words :
" Long haue they lien hidden in obfcuritie, and
happily [haplie] had neuer feene the light, had not
a meere accident conuayed them to my hands," etc. ;
but alfo, moft fortunately, the Title-page, for it
revealed the name of the illuftrious author.

I therefore fent a communication at the end of
October, 1873 (inferted November 1), to the
" Athenæum," which, from its high character and
world-wide circulation, was moft likely to effect my
object. Nor was I difappointed, for a few days
after I received the following note :

" St. Mary's College,<br>
" Ofcott, Birmingham.<br>
" Nov. 8, 1873.

" Dear Sir,

  " Would you kindly tell me whether the fragment of the poem of Southwell which you have
difcovered begins thus :

   'O wretched man which loveft earthlie thinges<br>
   And to this worlde haft made thyfelfe a thrall.'

   .   .   .   .   . .   .   .   .   .   .

  " This is the firft ftanza of a poem which we have
here at the Coll. in MS., and if I can identify it as

---

[1] I have always prefumed this " W. H." to be the fame
" W. H." who gave Shakefpeare's Sonnets to the world three
years after the prefent work was iffued from the prefs of the
fame printer, George Eld.

Southwell's I fhould think it worth while, with the Prefident's permiſſion, to have it printed. In any cafe, as a Catholic, I fhould wifh to thank you for bringing to light fomething illuftrative of the life and works of F. Robert Southwell, and therefore of fuch intereft to Englifh literature. Believe me, Dear Sir, yours very truly, S. SOLE.

"Charles Edmonds, Efq."

A few days later I received the following letter from the Prefident, who, after expreſſing his regret at not being able to fee me when I called owing to prefs of bufinefs, continues thus: "Mr. Sole has explained to me your wifh to publifh the whole of this poem of Southwell's ; and as you have been the means of identifying the poem as his, I think it is only fair that you fhould receive every help we can give you in carrying out your defire. I therefore will fend you the MS. tomorrow, trufting with confidence to your taking all poſſible care of it, and returning it to us as foon as you have tranfcribed this poem. Yours truly, J. SPENCER NORTHCOTE."

This was the title under which the " Fourefould Meditation " was concealed; probably for fufficient prudential reafons : "Sartaine mofte holfome & neceſſarie confiderations, or meditations verye meete and convenyent (for all degrees) and att all tymes to be duelye confidered of and had in Rememberance To withdrawe our affeċtions from this vaine & wicked worlde, to the defire of Heaven and heavenlye thinges. Reade with good advifement."

*b*

The volume confifts of 180 leaves, and at *the beginning of the MS. is this:* "The Epiftel Dedicatorie. To the right worfhipfull Mr. Thomas Knevett Efquire, Peter Mowle wifheth the perpetuytie of true felyfitie, the health of bodie and foule with continewance of worfhipp in this worlde, And after Death the participation of Heavenlie happines dewringe all worldes for ever." Among other pieces in the volume are :

"A brief Catachifm of Chriftian Doctrine, compyled by Lawrence Vaux, Bachelor of Divinitie, 1583." 41 *leaves.*

[Of the family of Baron Vaux of Harrowden, which title, created in 1524, is now extinct, but revived in the perfon of Lord Chancellor Brougham in 1830, whofe anceftor married Jane Vaux.]

Peter Mowle his Loking Glaffe.

Certaine of Alabafters his Meditations. Anno 1597. 13 *ftanzas of* 14 *lines each.*

Defiderius, or the readie way to the Love of God. Written in Dialogue wife, under learned and pleafaunt Allegories. Firft put forth in the Spanifhe tonge and after tranflated into Latin : and now lately into Englifhe for the behoofe of the devout of our nation by I. G. Prifoner. *In profe:* 28 *clofely-written leaves.*

[The famous FATHER JOHN GERARD, author of "The Narrative of the Gunpowder Plot, who fled with Southwell when purfued by four Prieft-hunters or purfuivants.]

Sartaine Godlye and devout Verefes of the paffion of our Lord and Savyor Jefu Chrift, the Lamentation of our bleffed Ladie (in Latin Stabat Mater dolorofa, &c.) the fiftene mifteries of the Rofarie of our Ladie in verfe, with dyverfe other godly prayers and devoute matters fett forth by S. W. and dedycated to the vertuous Ladie Pawlett.

The Difcourfe of the Martirdome of Mrs. Margarett Clytherowe ; A.D. 1586.

Verfes given for a New Yeares Gift in Anno 1592 to the Ladie Vifcountis Hereford of Parham.

Verſes of the Earthquake which happened on the 24th daie of December 1601.

The Anatomie of Pride made by mee P.M. 1602.

A devout and godly prayer made by the moſt excelent and godlye Queene, Queene Marye.

Verſes to The Worſhipfull my good mſs. Miſtres Elenor Woodhowſe of Caſtor. Anno 1606.

At end: "Peter Mowld, Junior, oweth this Booke. Wittneſſe Edmond Mould. 1605." While the witneſs calls himſelf Mould, the owner uſes indifferently the names Mowld and Mowldc. He deſcribes himſelf as of Attelbroughe, and of his being in 1589 in his 35th year.

The dated pieces range from 1590 to 1606.

The *Oſcott MS.* is *not* followed in the preſent reprint for the following reaſons: it contains only 118 ſtanzas, while that in the *Rawlinſon* collection in the Bodleian contains 126; the additional ones being Nos. 42 and 63 to 69. Not only is the *order* of ſtanzas 13 and 14 different, but they vary in the commencement of the former. And the *printed* fragment ſhows that the reading there given muſt have preceded that of the *Rawlinſon MS.* The latter is therefore uſed; but it contains no title-page, and is aſcribed erroneouſly to Lord Philip Arundel.

I find that Southwell has *Poems* in "Briefe Meditations in the moſt Holy Sacrament," by L. PINELLI, of the Society of Jeſus; alſo "Hymes [*ſic*] gathered out of S. Thomas de Aquino, tranſlated by the Rev. Fa: R. S." 8vo., *s. l. et a.*

On Tueſday, March 26th, the following intereſting MS. was ſold at Sotheby's. Lot 1050, Bibliotheca Phillippica. This MS. formerly belonged to the famous hagiographer, Alban Butler, whofe autograph appears upon the firft page.

" 1050   Southwell   or   Śotwell.    Meditationes Roberti Sotuelli Martyris de Attributis Divinis ad amorem Dei excitantes—Exercitia et Devotiones ejufdem, in the original vellum binding.    8vo."

C. E.

A

# FOVREFOVLD
## Meditation,

*Of the foure laſt things :*
*viz.*

$$\left.\begin{array}{l} 1. \\ 2. \\ 3. \\ 4. \end{array}\right\} \text{ of the } \left\{\begin{array}{l} \text{Houre of Death.} \\ \text{Day of Iudgement.} \\ \text{Paines of Hell.} \\ \text{Ioyes of Heauen.} \end{array}\right.$$

*Shewing the eſtate of the Eleƈt and Reprobate.*

Compoſed in a Diuine Poeme

By *R: S.*
*The author of S. Peters complaint.*

Imprinted at London by *G. Eld:* for *Francis Burton.*
1 6 0 6.

# To the Right Worſhipfull and
## *Vertuous Gentleman, Mathew*
### Saunders, Eſquire.
### W.H. wiſheth, with long life, a proſperous
### achieuement of his good deſires.

*Ir ; as I with great deſire apprehended the leaſt opportunity of manifeſting towards your worthy ſelfe my ſincere affeCtion, ſo ſhould I be very ſory to preſent any thing vnto you, wherein I ſhould growe ~ offenſiue, or willingly breed your leaſt moleſtation : but theſe meditations, being Diuine and Religious (& vpon mine owne knowledge, correſpondent to your zealous inclination) emboldened me to recommend them to your view and cenſure, and therein to make knowne mine owne entire affeCtion, and ſeruiceable loue towards you. Long haue they lien hidden in obſcuritie, and happily had neuer ſeene the light, had not a meere accident conuayed them to my hands. But, hauing ſeriouſly peruſed them, loath I was that any who are religiouſly affeCted, ſhould be depriued of ſo great a comfort, as the due conſideration thereof may bring vnto them. As for my ſelfe, Sir, the knowledge you haue of me, I hope will excuſe the coldneſſe and ſterilitie of my conceipts, who couet to illuſtrate my intire affeCtiō vnto your worſhip, by reall and approued aCtions, referring my ſelfe wholly in this, & all other my indeuours, to your fauourable conſtruCtion, who ſhall euer be of power, in the humbleſt ſeruices to command me.*

Your Worſhips vnfained affeCtionate
W. H.

# A Treatife of the houre of Death,

*the day of Iudgement, the paines*
of Hell, and the ioyes of Heauen.

---

## *Of the houre of Death.*

### 1.

O Wretched man, which loueft earthlie thinges,
　　And to this worlde haft made thyfelfe a thrall,
Whofe fhorte delightes eternall forrow bringes,
Whofe fweete in fhewe in trewth is bitter gall :
　　Whofe pleafures fade eare fcarfe they be poffeft,
　　And greve him left that moft doe them deteft.

### 2.

Thou arte not fuer one moment for to lyue,
And att thy death thou leaueft all behinde,
Thy landes and goodes noe fuckor then can geue,
Thie pleafures paft are croffes to thie minde :
　　Thie friend the world can yeld thee noe releefe,
　　Thy greateft ioye will proue thie greateft greefe.
　　　　　　　　　　　　　　　　　　The

## Of the houre

### 3.

The tyme will come when Death will thee aſſalte :
Conceyue yt then as preſent for to bee,
That thou in tyme maieſt ſeeke to mend thie falte,
And in thie life thine errors plainlye ſee :
   Imagen now thie corſe is allmoſt ſpent,
   And marke thie frinds how deepelie they lament.


### 4.

Thy wyfe dothe howle, and pearce the verie ſkies,
Thie chilldrens teares their ſorrowes doth bewraye,
Thie kinesfolke morne and wepe with woefull cryes,
Now thou muſt dye, and canſt noe longer ſtaye :
   Loe here the ioyes and treaſures of thie hart :
   Thie race is ronne : from them thou muſt depart.


### 5.

With paine thou doſt lye, gaſpinge all for breath,
Paſt hope of life or hope of anie good,
Thy preſent ſtate a lyuelye forme of death,
Thie hart become all cold for want of blood :
   Thie noſethrills ronne, and gaſpinge thou doſt lye,
   Thie lothſome ſight thie frinds beginne to flie.
                       Thy

## of Death.

### 6.

Thy voyce doth yeld a horce and hollowe founde,
Thie dyinge head doth greadie feeme to fleape,
Thie fences all with horror doth abound,
Thie feete doth die, and death doth vpward creepe:
  Thie eyes doth ftand, faft fett into thine head,
  Thie jawes doth fall, and fhowe thee allmoft dead.

### 7.

What dofte thou thinke, now all thie fences faile?
What dofte thou faye by pleafure here is wonne?
How doft thou now thie paffed life bewayle?
How doft thou wifhe thie courfe were new to ronne?
  What woldft thou doe thie endinge life to faue?
  What woldft thou geue for that thou canft not
[haue?

### 8.

Thy bodie now muft frome the foule departe,
Thie lands and goods another muft poffeffe,
Thie ioyes are paft on which thou fetft thine harte,
Thie paines to come noe creature can expreffe:
  Loe here the fruite and gaine of all thie finne,
  Thie Life muft end, and Death muft now beginne.
                                                Thy

## *Of the houre*

### 9.

Thy former faultes are fett before thine eyes,
And monftrous fhewes which feemd before fo fmall,
To fwallowe thee, Defpaire in fecrett lyes,
And all thie finnes with terror thee appall : [mone,
   With fcalldinge fighes they make thee now to
   And in thie foule with forrowe thou doft grone.

### 10.

Thou wayleft now the pleafinge of thie will,
Thie euill gott goods doth make thee fo lament, '
Thievaine delightes with anguifhe thee doth fill, '
Thie wantone tricks thie confcience doth torment :
   Thie fweeteft finnes doth bringe thee bitter fmarte,
   Thie heynous faultes oppreffe thie dyinge harte.

### 11.

With dreadfull feare they fhake thie dolefull mynd,
And bent to fight, with force they thee inclofe,
In worldlye helpe noe reffcue thou canft finde :
And ftandinge now amidft thie mortall foes,
   A thoufand deathes wold feeme a leffer paine
   Then this eftate in which thou doft remaine.
                           Noe

## of Death.

### 12.

Noe tonge, no penn, nor creature can bewraye,
Howe all thie finnes their feftred rancor fhowe,
Howe dreadfull fightes with forrowe thee difmaye,
Howe bluftringe ftormes of greefe beginne to blowe:
    Thie ioyes are gone, which were thie God before,
    Thie life is done and fhall returne noe more.

### 13.

What booteth it thie lewdnes to repent,
And leaue to finne when finne forfaketh thee?
What canft thou doe when all thie force is fpent?
Will then our Lord with this appeafed bee?
    Thie life thou ledft in feruice of his foe,
    And farueft him when life thou muft forgoe.

### 14.

Now heauen to win noe paines thou wouldft refufe,
Nor fpare thie goods to eafe thie woefull ftate,
Of all thie finnes thou doft thie felfe accufe,
And call for grace when callinge comes to late:
    For finne thou dideft while life and power did laft,
    And leaueft now, when force to finne is paft.

    Then

## *Of the houre*

### 15.

Then had I wift, with forrowe thou doft faie,
But after witts repentance euer breed,
The daye is come, thie debt thou now muft paie,
And yeld to death, when life thou moft fhalt neede :
   Thie breath is ftopt in twinclinge of an eye,
   Thie bodie dead in vglie forme doth lye.

### 16.

Thye carcaffe now like carrion menn doth fhonne,
Thie frends doe haft thie buryall to procuer,
Thie faruaunts feeke from thee awaye to ronne,
Thie lothfome ftench noe creature can induer :
   And they which tooke in thee their moft delight,
   Doe hate thee moft, and moft abhorre thie fight.

### 17.

Thye flefh fhall ferue for maggotts for a praye,
For pamperinge which both fea and land was fought,
Thie bodie muft tranceformed be to claye,
For whofe delight fuche coftlie clothes were bought :
   Thie pryde in duft, thie glorie in the graue,
   Thie flefh in earth their endinge now fhall haue.

Behold

## of Death.

### 18.

Behold! the place in which thou doſt abyde
Is lothſome, darke, vnſweete, and verie ſtraite :
With rotten bones beſett on euerye ſyde,
And crawlinge wormes to feede on thee doth waite :
   Oh harde exchange! O vile and hatefull place!
   Where earth and fillth thie carcaſe muſt imbrace.

### 19.

O wretched ſtate! O moſt vnhappie man!
Yet were yt well yf nothing were behinde,
Yf all myght end as here yt firſt begann,
Some comfort were ſuche endinge for to finde :
   For then as God of nothinge thee did frame,
   By courſe againe thou ſhouldſt become the ſame.

### 20.

But lyue thou muſt a thouſand deathes to die,
And dyinge ſtill, yet neuer whollie dead,
Thou muſt appere before the Judge on hie,
And haue reward as thou thie life haſt ledd :
   Thie tyme is come, thou canſt no longer ſtay,
   The iudge is ſett, and boteleſſe is delaye.
                      Behoulde

# *Of the day*

### 21.

Behoulde his power.   Loe whom thou didſt offend
For vaine delights, which were but mere deceipt,
Behould on him how Anngells doth attend,
And all that court doe for his comminge waight :
  Behould his throne of glorie in the ſkies,
  And ſee how wrath doth ſparkell from his eies.

### 22.

Loe this is hee whoe euerie thing did make, [daye,
Whom Heauen and Earth doe prayſe both night and
Loe here the looke att which the Anngells quake,
Loe here the Lord whom all thinges doth obaye :
  His will is lawe, and maye not be withſtand,
  His wrath conſumes and killeth out of hand.

### 23.

O filthie foule, how maye this wrath be borne ?
Or can a worme his furie now abyde ?
The Anngells laugh thy fillthines to ſkorne :
They hate thie ſinne, and thee for ſwellinge pryde :
  They ſhine with beames fare brightter then the
  And call on God that Juſtice may be done. [Sonne,
                                                    Each

## *of Judgement.*

### 24.

Each creature cryes that punifht thou mayft bee,
Whom in thie lyfe thou lewdlye didft abufe :
Both Heauen and earth are fooes proteft to thee,
And all thie thoughtes of finne doth thee accufe :
   Thie wordes and deedes againft thee now are
       brought,
   And all thie filth which finne in thee hath wrought.

### 25.

Thou fyted arte a juft account to fhowe,
How farre thou fought thie felfe for to deny,
How all thie landes and welth thou didft beftowe,
And with thie goodes thie brothers wante fupplye :
   What care thou hadft thie makers name to prayfe,
   What paine thou tokft to walk in all his wayes.

### 26.

The Judge dothe afke how all thie life was fpent,
Yf from offence thie fences thou didft keepe,
Yf in thie foule thou truelye didft repent,
And for thie finne with hartie forrowe weepe :
   Yf thou his feare didft fett before thine eyes,
   And for his loue all worldlie ioyes defpife.

Yf

## Of the day

**27.**

Yf eke thie foes reuenge thou hafte not wrought,
Yf to thie frindes thou neuer wert vnkinde,
Yf earthlie pompe thou euer fett att nought,
Yf fecrett hate thou hafte not kept in mynde:
   Yf thou alike didft ioye and forrowe take,
   And with thie harte all carnall luft forfake.


**28.**

Thye thoughtes and wordes the Judge dothe open
And afketh now a ftrayte account of all,     [laye,
How thou didft here his motions obaye,
And for his grace with erenaft fervor call:
   Yf all thie lyfe on earth thou ledft vpright,
   And in his loue didft fett thie whole delight.


**29.**

What canft thou plead thie lewdnes to excufe,
When truth fhall proue in all thou didft offend?
The Judge is juft, thou mayft not him refufe,
Thie caufe is naught, thou canft not it defend:
   To hope for helpe, alas! it is in vaine,
   The tyme is pafte, noe helpe thou canft obtaine.
                           Our

## of Judgement.

### 30.

Our Lord doth faye, " how couldſt thou uſe me foe,
Sith I to thee both foule and bodie gaue?
How durſt thou feeke and ferue my mortall fooe,
Sithe I did dye thie felfe from death to faue?
   I gaue thee all, and me thou didſt deteſt,
   He gaue thee naught, yet wholie thee poſeſt.

### 31.

" Thye lands and life did from my goodnes flowe,
Thy fleſhe and bones I did of nothinge frame,
Both wellth and witt I did on thee beſtowe,
And gaue thee all to prayſe my holie name:
   Yett with them all againſt mee thou didſt fight,
   And fledd to them whoe bredd mee greateſt ſpight.

### 32.

" When I did ſpeake thou ſeemedſt deafe and dombe,
When he did call thou madſt him aunſwere ſtrayte,
He neuer ſtayd but thou didſt quickly come,
And I without inforſed was to wayte:
   O thankeleſſe wretche thou mee ſhalt ſee noe more,
   But dwell with him whoe had thie harte before.
                          Thou

## *Of the day*

### 33.

" Thou fhalt with him for euer more remayne,
To whome thie felfe for pleafure thou hafte foulde,
His will thou wroughft, and myne thou didft dif-
His right thou arte, I can not thee withoulde: [daine,
　Thie owne deferts haue made thee his to bee,
　The choyfe was thine, noe wronge is donne to
　　　　　　　　　　　　　　　　　[thee."

### 34.

Then comes the Devill, and to our Lord doth faye,
" O righteoufs Judge, this wretche I ought to haue,
For in his lyfe he would not thee obaye,
But with his harte to mee him felfe he gaue:
　My precepts eke he practift daye and night,
　And mee to pleafe he made his whole delight.

### 35.

" Him felfe he vowed to ferue me all his dayes,
His eyes were fixt vppon my counfell ftill,
His feete were bent to walke in all my wayes,
His harte was fett for to performe my will:
　His life and landes I drue him on to fpend,
　In doeinge that which might thee moft offend.
　　　　　　　　　　　　　　　　　Hee

## *of Judgement.*

### 36.

"Hee fcornd thie power and quyte refufde thie grace,
Thie bitter paynes hee bannifht from his eyes,
Thie precious bloud hee never would imbrace,
Thie gracious woundes he lewdlie did defpife :
    Thie threats for finne he reckoned as a ieft,
    Thie wordes and will in all he did deteft.

### 37.

"Thie glorious death hee feemed to difdaine,
And followed that in which hee did delight,
For fervinge thee he toke not anie paine,
But all thie love with hate he did requite :
    What reafon then thie glorie he fhould fee,
    Of which he feemde fo carelefTe for to bee.

### 38.

"Thou didft him make, and on him all beftowe,
I nothinge gaue nor him to beinge brought,
Yet thee he left, to whom he loue did owe,
And mee hee farvd, whoe never gave him ought :
    What woldft thou more thou vfeft not to wronge,
    And hee to mee in Juftice doth belonge.

D            Behoulde

## Of the day

### 39.

Behoulde, O foule! how God doth thee refufe,
And how his foe doth clayme thee as his owne,
Thie confcience doth with horror thee accufe,
And reape thou muft as thou before haft fowne :
    The Lord of Lords doth thee condemne to lye
    In endleffe flames where livinge thou fhalt dye.

### 40.

O wretched foule! what fhall become of thee ?
What greater paine can any harte devife ?
Yett worfe their is, if worfe their yett maye bee,
Thie bodie muft to Judgment fhortlie rife :
    And bothe alike in Hell muft fuffer fmarte,
    As both in earth in finne had equall parte.

### 41.

All finners faine would fhonne this dreadfull daye,
And wifhe yt were without their perill paft,
The feare alone muft needs their hartes difmaye,
The fignes appeare and on yt cometh faft :
    Behold the Sonn is darke which fhined bright,
    The ftares doe fall, the moone hathe loft her light.
                         Behould

## of Judgement.

### 42.

Behould how men are witherede quite with woe,
And cannot find a harbowre now of reſt :
Behould on earth how fencleſſe they doe goe,
Theire faces palle, theire harts with feare oppreſt :
   Behould each where how beaſts for terrour cry,
   And marke how men alredy feeme to dye.

### 43.

Behoulde how blodd the trees and braunches fweate,
And howe each thinge in trembblinge wiſe doth
Behoulde the Sea againſt the Land doth beate, [quake,
And roringe lowde doth force the Earth to ſhake :
   Her ſurges mounte, her ſwellinge furie ſhowes,
   And on the Land her fiſhe with rage ſhee throwes.

### 44.

The clowdes like ſmoake doe thicken in the ſkies,
The mountaines move, the Earth doth open wide,
The bluſtringe windes with ſtormes and tempeſts
The ſtowtteſt hartes their faces feeke to hide : [riſe
   Both ritch and poore from citties now are fledd,
   And all in caves doe ronne to ſhrowde their head.
                         Eche

## Of the day

### 45.

Eche lyvinge thinge for helpe doth crye and call,
And favage beaftes vnto the Cittie flie,
'The earth doth quake, the ftrongeft towers fall,
And beaftes remaine were menn did vfe to lie :
    The courfe begins of nature heire to faile,
    The Heauens doth mourne and all thinges els
                     [doth wayle.

### 46.

The Anngells lowd their Trumpets dreadfull found,
And fummones all that ever lyfe pofeft,
The Earth with woe and terror doth abound,
The dead aryfe that longe had bene at reft :
    Bothe quicke and dead affembled round doe ftand,
    And wayte his will whofe comminge is at hand.

### 47.

Behouldè how lowe both Heaven and earth doe bowe,
And proftrate all his favor to defyre,
Behould howe Chrift in glorie cometh now,
And in the ayre appeares a flame of fyer :
    The Earth for feare doe tremble att this fight,
    The fea is dryed, the hills are molten quight.
                            The

## of Judgement.

### 48.

The hardeft rockes are turned into duft,
His furious wrath noe creature can abyde,
Their paines were fweete which now are proved juft,
And neede not feeke in corners them to hyde:
   Our Lord rewardes as merytt hee doth finde,
   Thrife happie they that beare a giltles minde.

### 49.

O curfed foule! how art thou drownd in care,
When all this fight is fett before thine eyes:
Thy paffinge feare noe wrytinge can declare,
Thie bodie darke like Deathe doe feme to ryfe:
   Thie hope is paft for eafinge of thie fmarte,
   Thie finnes are prickes to wound thie dyinge
                           [harte.

### 50.

Behould how thou noe favor here canft gett,
Nor from thie foes by anie meanes efcape:
Thie right hand is with all thie finnes befett,
Beneath thee Hell to fwallowe thee doe gape:
   The fearefull fends vppon thie left hand frowne,
   And lye in wayte, to throwe thee hedlonge downe.
                             Above

## Of the day

51.

Above thee fytts the Judge all fild with rage,
Whom in thie life thou lewdlie didft offend,
Noe helpe thou haft his furie to affwage,
His browes hee doth with anger fercelie bend :
 And all the finnes of menn hee doth repeate,
 Which forceth now his furie to be greate.


52.

Within thee gnawes thie confcience voyde of grace,
And all the evill to which thou didft confent,
Without thee ftands thie frinds which wayle thie cace,
And doe thie ftate with bitter grefe lament ;
 On euerie fyde the world doth thee affright,
 Whofe terror fhowes, with flames that burneth
 [bright.

53.

If forward now thou tookeft on thie waye,
Thou hedlonge doft vnto thie ruine run,
The devills doe watche thie goinge backe to ftaye,
Noe meanes is left misfortune for to fhun :
 What wilt thou doe, invirond thus with woe ?
 For neyther back nor forward thou canft goe.

O

## *of Judgement.*

### 54.

O wretched man ! how heauie is thie harte,
How doſt thou wiſh for that which can not bee,
How doſt thou ſigh and quake in euerie parte,
And muſt thie frinds be ſeuerd thus from thee :
    They fild with ioye in glorie now ſhall raigne,
    And full of greife thou torment muſt ſuſtaine.

### 55.

The Judges wordes are like a burninge fyer,
Which waſteth all it commeth to imbrace,
It booteth not his mercie to requyer,
The time is paſt of callinge now for grace :
    Behould the Judge doth thee condemne to hell,
    Wher thou in paine for ſinne ſhalt ever dwell.

### 56.

O dolefull wordes ! O moſt vnhappie wight !
Thie head to ſhrowd for mountaines thou doſt call,
Thie future paines are preſent in thie ſight,
And curſeſt now the cawſes of thie fall :
    Thie birth and life to late thou doſt repent,
    Yet wayleſt both and doſt in vaine lament.

What

## Of the paines

### 57.

What tonge, what penn, what creature can expreſſe
Thoſe deadlie greifes which allwayes thou doſt taſt?
The longer tyme the comfort is the leſſe,
Thie hope decayes, thie ſorrowes never waſt.
  O bitter ſweete that earthlie pleaſures breede !
  Thie livinge death all torments doth exceede.

### 58.

Thye wanton eies thoſe helliſh monſters ſee,
Whoſe blodie mindes thie ruine did conſpire,
Whoſe neeſinge ſeme like lightning for to bee, [fire :
Whoſe monſtrous mouthes doe caſt out flames of
  Whoſe noſethrills ſmoake, whoſe eies are glowing
    redd,
  Whoſe whole delight by others ſmarte is bredd.

### 59.

Thye wrètched eares, which harkened vnto lyes,
Doe here howe fends doe rage with all deſpight,
Noe noyſe is their but ſhreekes and hideous cryes,
Which able are the ſtouteſt hart to fright : [wayle,
  Wher ſome blaſpheme, and ſome their ſtates be-
  Where others curſe and never ceaſe to rayle.
                     Thye

## of Hell.

### 60.

Thye daintie noſe, which had perfumes ech daye,
A lothſome ſtenche for ever muſt abyde,
Which riſeth vpp from dampned bodies aye,
That heaped their doe lye on euerie ſyde :
    Loe here the ſweete thie ſmellinge to content,
    Noe worldlie filth can yeld ſo fowle a ſent.

### 61.

Thye curyous taſt doth hunger their ſuſtaine,
Which did in meates ſuch rare deviſes crave,
With burninge thirſt thou ſuffreſt grevous paine,
And yt to coole noe water thou canſt haue :
    ·Noe dropp is their, thie thirſtinge for to eaſe,
    Noe hope of helpe that maye thie grefe appeaſe.

### 62.

Thye feelinge yett the greateſt paine doth beare :
With fierie flames which all thie partes torment,
An extreame cowld thou allſo findeſt their,
With gnaſhing teeth that makes thee to lament :
    Thie teares with heat in ſtreames are daylie ſhedd,
    Thie teeth for cowld doe chatter in thie hedd.

E               If

## Of the paines

### 63.

If for a while noe creature can endure
In earthly fiere one member for to bee,
What torments doe thy paſſed Joyes procure,
In endleſſe flames thy members all to fee !   [breed,
   What greefe, what paine, what ſorrowes doe they
   Which earthly flames in all doe farre exceede !

### 64.

The deiuills with flouts doe lough the now to ſcorne,
Thy fleſh and bones in ſunder they doe teare,
Thy curſed ſkinne with cruell whipes is worne,
Thy woefull harte is filled full with feare :
   With inwarde woe thy ſoule is ſore oppreſte,
   With outward paine thy body finds no reſte.

### 65.

Thy torments ſtrange doe breede thee bitter greefe,
And reſte in thine Imagination ſtill,
Thyne owne conceipte which now ſhould yeld releefe,
Doth labour more with ſorrow thee to fill :   [chew,
   Thou thinkeſt moſt what moſt thou whouldſt eſ-
   Thy griefe thy thoughts, and thoughts thy griefe
       renew.                 Thy

## of Hell.

### 66.

Thy memory doth call vnto thy mynde
The ſhorte delight of all thy pleaſures paſt,
Yt wounds thy harte theſe paines for them to finde,
Which greueous are and ſhall for euer laſt :
    Thy deſperate caſe no comfort can obtaine,
    Thy paſſed Joyes encreaſe thy preſẽt paine.

### 67.

Thine vnderſtandinge doth thy miſery ſhew,
And telleth thee thou arte in Sathans Jawes,
For ſhorte delights, thy loſſe yt makes thee know,
And in thy ſoule the worme of Conſcience gnawes :
    Thoſe fadinge Joyes in rage thou doſt defye,
    And in diſpight they make thee thus to crye.

### 68.

" My former Joy a ſhadow was in deede,
It did not laſt, but paſſed quicke away,
My preſent paine all meaſure doth exceede,
Noe witt nor arte my torments can bewray :
    A time there was when bliſſe I might haue woone,
    But time is paſt, and all my courſe is runne.

O

## Of the paines

### 69.

" O curfed time, in which I time forfooke,
A litle paine had ridd me of my woe !
O curfed Joyes in which I pleafure tooke,
For pleafinge you all pleafures I forgoe !
   And here in hell each kinde of paine I finde,
   Which wafts my fleafh and wounds my woefull
                     mynde.

### 70.

" Yf I my finnes with forrowe had confeft,
They had to me bene clene remitted all :
In ftead of greefe, I glorie had poffeft,
If I for grace had bent my minde to call :
   O wretched wretch, that for fo fmall a paine,
   Refufinge bliffe, in torment muft remaine.

### 71.

" The greateft ioyes which doe in earth abound
Can in a world not yeld fo much delight
As here by paine is in a moment found,
Whofe blafinge woe is prefent ftill in fight :
   What fancie then bewitched my wretched harte,
   For fained Joyes to fuffer endleffe fmarte.

My

## of Hell.

### 72.

" My parents were the cawfers of my woe,
And all the meate on which I euer fedd,
My carnall frind hath proued my greateft foe,
And vnto mee this mifchefe now hath bredd :
 Accufe mee all that hathe my ruine wrought,
 And euerie meane which mee to beinge brought.

### 73.

" Thrife happie they on earthe that never were !
Their ftate is bleft that never came to liue !
O bleffed wombes that chilldren never bare !
O happie breft which fuck did never geve !
 O deadlie paine ! O moft unhappie place !
 O curfed wretch whome ill mifhapps imbrace !"

### 74.

Loe here the plaints in this infernall lake,
Wher Scorpions ftinge and fquorges thee torment,
Wher hammers beate, and Devils a roringe make,
Wher hope is paft and dampned foules lament :
 Wher wormes doe crawle and uglie ferpents creepe,
 Wher paines abound, and forrowes make thee
  weepe.        Againft

## Of the paines

### 75.

Againſt our Lord thou raieſt with deſpight,
And him thou doſt with raginge words defie,
Thou barred art from ſeeinge anie light,
And while ye liue thou muſt for ever die :
  Loe here the fruite which worldlie pleaſures bringe,
  Thie paines agree in meaſure with thie ſinne.

### 76.

Thye ſweet delights are come to woe and wrack,
Thie happie ſtate unto a wretched caſe,
Thie gredie minde is punniſht here with lack,
Thie lecherous armes doe uglie fends imbrace :
  Thie envious ſowle doth howle for deadlie paine,
  Thie haughtie harte doth ſuffer depe diſdaine.

### 77.

Thou findeſt ſmart in ſtead of pleaſaunt games,
Thie daintie wynes are turnd to bitter gall,
Thie coſtlie clothes are now made burning flames,
Thie loftie pride hath now a lothſome fall :
  Thou nothinge doſt which maye afford thee eaſe,
  But feeleſt all which maye thee moſt diſpleaſe.

                                              Yet

## *of Hell.*

### 78.

Yet cheiflie one which farre doth all exceade,
And as it is none rightlie can efteme,
It greves thee moft and makes thie harte to bleed,
And joynd with it the other nothinge feeme :
 Then judge what paine this torture brings to thee,
 When matche to it all nothinge femes to bee.

### 79.

Thye fcences feele for everie finne a paine,
So rated their as here thou tokft delight,
And now for that our Lord doth thee difdaine,
Thou bannifht art for ever from his fight :
 The paine of fcence fmall torment thou doft finde,
 When thou this loffe doft call unto thie minde.

### 80.

A grevious loffe which cannot be expreft !
O caufe of greife and fpringe of deadlie woe,
The Soule hath loft the center of her reft,
Thie hope, thie helpe, thie life thou muft forgoe :
 Noe paine or loffe with this maye be comparde,
 It paffeth all and cannot be declared.

From

## *Of the paines*

### 81.

From hope of joye this is an endleſſe barr,
And greateſt plague that God on ſinn beſtowes:
Compard with this thy tortures pleaſaunt are,
And all thie loſſe an eaſie burthen ſhowes:
　　Thie bittreſt paines are trifles in thine eyes,
　　Thie burninge flames thou ſeemeſt to deſpiſe.

### 82.

What woe, what ſmarte, what paine can be expreſt,
Which wayteth now on thee for to be layde!
With ſwordes of greefe thie harte is daylye preſt,
With dreadfull feare thie ſcences are diſmayde:
　　Thie eie hath loſt what moſt ſhe did deſire,
　　Thie bodie burnes in flames of endleſſe fire.

### 83.

And yf thie paines an endinge might obtaine,
When yeres their were of manie thouſandes runn,
As on the earthe have lightten dropps of rayne,
Since firſt of all this wretched world begunn: [minde,
　　Some helpe this hope might bringe unto thie
　　When hope were left an end at laſt to finde.
　　　　　　　　　　　　　　　　　　　　But

## of Hell.

### 84

But of them all noe eafe nor end thou haft,
Within thie foule fome comforte might procure :
Noe tyme will helpe thie forrowes for to wafte,
While God is God thie torture fhall indure :
  Thie paine in truth is more then can be tould,
  The fight in thought noe creature can unfould.

### 85.

O dyinge lyfe ! O fea of endleffe fmarte !
Which nature hates and all thinges ellfe deteft,
O lyvinge death, noe life or death thou arte,
For death hath end and life hathe fometyme reft :
  The worft of both our Lord hath put in thee
  That neyther reft nor end might other bee.

### 86.

O dampned foule ! howe doft thou roare and crye !
What deadlie greefes thee daylie doe oppreffe !
But lyft a whyle thie curfed eies on hye,
And fee what ioyes the bleffed their poffeffe :
  That by the fight, thie torments maye increafe,
  And for thie loffe thie forrowes neuer ceafe.

F                                    And

## *Of the ioyes*

### 87.

And firſt behould the beawtie of the place,
Wher all the Saintes with Chriſt in glorie raigne,
Wher honor is not mixed with diſgrace,
Wher ioye is free from taſk of anie paine :
   Wher great rewards attend on good deſarts,
   And all delightes poſeſſeth faithfull harts.

### 88.

O wicked wretche ! This cittie now behould,
Which doth ſurppaſſe the reache of anie thought,
The gates are pearle, the ſtreetes are fyneſt gould,
With precious ſtones the walles are wholie wrought :
   Of Sunn and Moone it needeth not the light,
   For ever their the Lambe is ſhining bright.

### 89.

And from His ſeate a chriſtall river flowes,
Wher life doth runn, and pleaſures ever ſpringes :
On everye ſyde a tree of comforte growes,
Which ſavinge helthe to everie nation bringes :
   It worketh reſt, and ſtinteth worldlie ſtryfe,
   It flieth death, and bringeth endleſſe life.

This

## of Heauen.

### 90.

This goodlie place all beawtie doth furmount,
And all this world in largeneſſe paſſeth farr:
The earth it ſelfe in bignes in account
Not equall is unto the ſmalleſt ſtarr:
O worthie place whoſe glorie doth excell!
Thriſe happie they that their attaines to dwell!

### 91.

Noe Sainte their is but brighter ſeemes to bee
Then Sunn or moone whoſe beawties wonders breede:
What glorie then ſo manie Saintes to ſee,
Which all the ſtarrs in number farr excede!
All glorious their wher glorie doth abound,
O bleſſed ſtate wher bliſſe is ever found!

### 92.

Archangells are but underſarvaunts there,
And Anngells doe their makers will obaye,
The powers in ioye with triumpth doe appere,
The beawties ſhine, the thrones their beames diſplaye:
The Cherubins doe yeld a famous light,
The Seraphins with love are burninge ſhininge
bright.                                    Here

## Of the ioyes

### 93.

Here Patriarkes haue their ioye for all their paine,
The Prophets eke with endleſſe glorie bleſt,
The Martirs doe a worthie crowne obtaine,
The Virgins finde a hauen of happie reſt :
    To all their ioyes in glorie they are mett,
    And now poſeſſe what longe they ſought to gett.

### 94.

Thoſe ſacred Saintes remaine in perfeƈt peace,
Which Chriſt confeſt and walked in his wayes,
They ſwim in bliſſe which now ſhall never ceace,
And ſinginge all, his name for ever prayſe :
    Before his throne in white they daylie ſtand,
    And carrie palmes of triumpth in their handes.

### 95.

The Angells then are next in their degree,
Whoſe order is in number to be nyne,
Noe harte can think what ioye it is to ſee
Howe all thoſe troupes with lampes in glorie ſhine :
    The ioye is more then wrytinge can expreſſe :
    O happie eies that maye theſe ioyes poſeſſe !
                     Above

### 96.

Above them all the Viregin hath a place,
Which cawfd the world with comfort to abound :
The beames doe fhine in her unfpotted face,
And with the ftarres her head is richlye crownd :
    In glory fhee all creatures paffeth farr :
    The moone her fhooes, the funn her garments are.

### 97.

O Queene of Heauen ! o pure and glorious fight !
Moft bleffed thou above all womenn arte !
This cittie druncke thou makeft with delight,
And with thie beames reioyfeft everie harte :
    Our bliffe was loft and yt thou didft reftore,
    The Anngells all and menn doe thee adore.

### 98.

Loe ! here the looke which Anngells doe admire !
Loe ! here the fpringe from whom all goodnes flowes !
Loe ! here the fight that menn and Saintes defire !
Loe ! here the ftalks on which our comfort growes !
    Loe this is fhee whom heaven and earth imbrace,
    Whom God did choofe and filled full of grace.

And

## Of the ioyes

99.

And next to her, but in a higher throne,
Our Saviour in his manhode fitteth here :
From whom proceedes all perfect ioye alone,
And in whofe face all glorie doth appere :
   The Saintes delight conceyved cannot bee,
   When they a man the Lord of Anngells fee.

100.

They ravifhed are with ioye in feeinge this,
How Chrift our Lord the higheft place obtaines :
They now behould the feate of endleffe bliffe,
And ioye to marke how hee in triumpth raynes :
   What ioye to menn moreover can befall
   Then here to fee a man the Lord of all ?

101.

More ioye yt yeldes then anie can devife,
A greater bliffe then may in words be tould,
His perfinge beames doth dazell all their eies,
His brightnes fcharce his Anngells can behould :
   The Saintes in him their wifhed comfort finds,
   And now inioye what moft content their minds.

To

## of Heauen.

### 102.

To thinke on this yt paſſeth humaine witt:
The more we thinke the leſſe we come to knowe:
He dothe uppon his Fathers right hand ſitt,
And all ye Saintes their humble ſarvice ſhowe:
    His ſight to them doth endleſſe comfort bringe,
    And they to him all prayſes euer ſinge.

### 103.

O worthie place, wher ſuche a Lord is cheife!
O glorious Lord, which princelye ſarvaunts keepes!
O happie Saintes, which never taſt of greife!
O bleſſed ſtate, wher malice ever ſleepes!
    Noe one is here of baſe or meane degree,
    But all are knowne the ſonns of God to bee.

### 104.

What higher place can anye prince attaine,
Then ſonne to him which ruleth all above?
Yet is their ſtate not ſubiect to diſdaine,
But in their mindes like brethren they doe love:
    Noe place is left for anie hate, or feare,
    But here they all one harte and ſoule doe beare.

O

## *Of the ioyes*

### 105.

O happie place, wher difcord never fights!
The ioyes of all are found in everie breft,
For ech as much in others ioye delights,
As if alone it in him felfe did reft:
   In all their ioyes noe difference is their knowne,
   For ech accounts them all to be his owne.

### 106.

And thofe they taft wherwith their Lord abounds:
As parte of theirs his glorie doe they take,
Unto them felues by union it redownds,
And all his ioyes their glorie perfect make:
   So fafte are knitt the members to the head,
   As over them his ioyes are whollie fpredd.

### 107.

What ioye is beft which here they doe not finde?
What greater bliffe, what pleafure maye be more?
What can by us conceyved be in minde
Which hath not bene recited here before?
   Yet one delight behinde as yet remaines,
   Which all in all, and all in it containes.

They

## of Heauen.

### 108.

They face to face doe God Almightie ſee!
And all in him as in a perfect glaſſe:
Noe good their is, but their is found to bee,
And all delightes this viſion doth ſurpaſſe.
　Ech ſight doth yeld the hart her perfect reſt,
　Becauſe noe good without him is poſſeſt.

### 109.

Hee preſent, paſt, and future thinges doth ſhewe,
And theirfore reſts their underſtandinge here:
Their nothinge is but they in him doe knowe,
And to their eies all plainlye doth appere:
　They now obteyne what longe they ſought to gett,
　And all their thoughtes are on him wholie ſett.

### 110.

Their will doth laſt in lovinge of his ſight,
In which conſiſts all good that cann be thought,
Shee here hathe fixt her love and whole delight,
And never will from lovinge this be brought:
　For here all good and goodnes doth abound,
　And never can without this good be found.

G　　　　　　　　Their

## *Of the ioyes*

### 111.

Their whole defire from hence doth never parte,
But fetled here for ever doth abyde :
This fight doth fill the mouth of everie harte,
And nothing leaves for them to wifhe befyde :
 Without defire, content fhee ftill remaines,
 And her defire with full delight obtaines.

### 112.

Their Faith behouldes her beft beloved gueft,
And her beleefe this fight doth here fullfill :
Their conftante Hope her hope hath now pofeft,
And him inioyes for whom fhee hoped ftill :
 Their Charitie, not perfect full before,
 To perfect ftate this vifion doth reftore.

### 113.

O glorious fight ! O fome of endleffe bliffe !
Which never wanes, nor feemeth for to wafte :
Whoe ever fawe foe fayer a fight as this,
Whoe ever did fuche heapes of comfort tafte ?
 What can be thought that can not here be hadd ?
 Where all doe ioye, and none are euer fadd.
They

## of Heauen.

### 114.

They here poſſeſſe what maye content them moſt,
And nothinge wante that perfect bliſſe maye bringe :
With all delight here breathes the Holye Ghoſt,
Which allwayes makes a freſhe and endleſſe ſpringe :
    Noe daye is here, noe morninge, noone, nor night,
    But ever one and allwayes ſhininge bright.

### 115.

O bleſſed ioyes, which all the ſoules poſeſſe !
O happie fruite, that vertue here hath wonne !
And in degrees the bodies finde noe leſſe,
But ſhine with beames farr brighter then the ſunn :
    Not ſubiect now to ſicknes, greife, or paine,
    · But glorious all, immortall they remaine.

### 116.

And propper ioyes ech ſence in private fyndes :
Their eyes behould that paſſinge glorious ſight,
Wher nothinge wantes for to content their mindes,
And all thinges elce which maye them moſt delight :
    Their eares are fedd with hearinge of ſweete ſoundes,
    And them to pleaſe all muſick here aboundes.

From

## *Of the ioyes*

### 117.

From fonges of praife the Saintes noe moment fpare :
Noe teares are feene nor anie their doe weepe :
But in this place the mufick is fo rare
As halfe a found would bringe all hartes a fleepe :
 And everie fence a propper pleafure takes,
 Which ioynd in one, their glorie perfect makes.

### 118.

Noe eie hath feene what ioyes the Saintes obtaine,
Nor eare hath hard what comforts are pofeft :
Noe harte can thinke in what delight they raigne,
Nor penn expreffe their happie porte of reft,
 Wher pleafure flowes, and greife is never fene,
 Wher good abounds, and ill is bannifht cleane.

### 119.

And of thofe ioyes noe creature end fhall fee :
The longer tyme the fweeter they doe fhowe :
While God indures they can not ended bee,
And never wafte, but allwayfe feeme to growe :
 When worldes are worne, and millions manie pafte,
 They now begin and fhall for ever laft.

O

## of Heauen.

### 120.

O ftate of ioye, wher endleffe ioye remaines!
O haven of bliffe, wher none doth fuffer wrack!
O happie howfe, which all delight containes!
O bleffed ftate, which never feeleth lack!
   O goodlie tree, which fruite dothe ever beare!
   O quyett ftate, which dannger neede not feare!

### 121.

O mixture pure, which bafeft droffe refynes!
O pleafaunte place, which onlie comforte bringes!
O ioyefull funn, wher glorie ever fhines!
O fruitfull foyle, wher pleafure ever fpringes!
   O glorious foules! O bodies wholie bleft!
   O fea of good, and of all good the beft!

### 122.

O dampned wretch! the thought of this alone
Oppreffeth thee with heapes of deadlie care,
And fighinge now in fperitt thou doft grone,
When with their bliffe thie woe thou doft compare:
   Thie greevous loffe dothe greive thie wretched
      harte,
   And yt with greefe redoubles all thie fmarte.

                                 If

## Of the ioyes

### 123.

If all the world by conqueſt thou hadſt wonne,
A trifle now thou thinkeſt all to geve,
That on the earth thie race were new to runn,
And thou againe wert ſuffered here to lyve :
    Another courſe thou wouldſt reſolve to take,
    And ſarvinge God thie carnall will forſake.

### 124.

The ſtraighteſt life thou woldſt noe paine eſteme,
Thie prayinge wold a paſſinge ioye appere,
Thie faſtinge ofte noe troble then would ſeme,
Nor anie greife the hardeſt penaunce here :
    A ioye thou woldſt account the ſharpeſt paine,
    To ſcape from Hell and endleſſe bliſſe obtaine.

### 125.

Now muſt I call, O worldlie man ! to thee,
The end wher firſt I did begin to wrighte,
That all theſe ioyes and paines which thou doſt ſee
May move thie minde to leade thie lyfe upright :
    Thie harte will melt to thinke uppon thie caſe,
    If their be left but halfe a ſparke of graſe.

Thou

## of Heauen.

### 126.

Thou findeſt here what thou wilt wiſhe att laſt,
And that account which none can ever ſhunn:
Then frame thie life before thie tyme be paſt,
As thou wilt wiſhe that thou in tyme hadſt donne:
    Leſt thou in vaine doſt waile thie wretched ſtate,
    When tyme is paſt and waylinge comes too late.